THIS GRAPH PAPER NOTEBOOK BELONGS TO:

____

____

About Brickshub

Brickshub is a design company based in England. We specialize in custom apparel, personalized products, and branded merchandise. We understand that we're not just designing shirts, jewelry, footwear, watches, journals, tote bags, drinkware, or home accessories; we are helping people make memories. They're keepsakes that will remain with you for years to come.

Thank you for being our customer and for allowing us the opportunity to be of service.

Made in the USA
Monee, IL
07 July 2026

56551522R00057